Exploring Food Chains and Food Webs

TIDE POOL
FOOD CHAINS

Katie Kawa

PowerKiDS press.

New York

Published in 2015 by The Rosen Publishing Group, Inc.
29 East 21st Street, New York, NY 10010

First Edition

Editor: Katie Kawa
Book Design: Reann Nye

Photo Credits: Cover Doug Cannell/E+/Getty Images; p. 5 (tide pool) David Mantel/E+/Getty Images; pp. 5, 21 (red algae) Vilainecrevette/Shutterstock.com; pp. 5, 21 (oysters) Jane Rix/Shutterstock.com; pp. 5, 21 (sea star) Logan Carter/Shutterstock.com; pp. 5, 21 (herring gull) iriksavrasik/Shutterstock.com; p. 6 De Agostini Picture Library/De Agostini Picture Library/Getty Images; pp. 7, 15 Ethan Daniels/Shutterstock.com; p. 8 Andrew J Martinez/Photo Researchers/Getty Images; p. 9 James Wheeler/Shutterstock.com; p. 10 Marevision/age fotostock/Getty Images; p. 11 Dudarev Mikhail/Shutterstock.com; pp. 12, 21 (sea urchin) NatalieJean/Shutterstock.com; pp. 13, 21 (limpets) Jon Boyes/Getty Images; pp. 17, 21 (hermit crab) wandee007/Shutterstock.com; pp. 18, 21 (bacteria) irur/Shutterstock.com; pp. 19, 21 (sponge) C.K.Ma/Shutterstock.com; p. 21 (tide pool) Steve Clancy Photography/Moment/Getty Images; p. 22 © iStockphoto.com/omgimages.

Library of Congress Cataloging-in-Publication Data

Kawa, Katie.
Tide pool food chains / by Katie Kawa.
p. cm. — (Exploring food chains and food webs)
Includes index.
ISBN 978-1-4994-0209-4 (pbk.)
ISBN 978-1-4994-0212-4 (6-pack)
ISBN 978-1-4994-0206-3 (library binding)
1. Tide pool ecology — Juvenile literature. 2. Food chains (Ecology) — Juvenile literature. I. Kawa, Katie. II. Title.
QH541.5.S35 K39 2015
577.69—d23

Manufactured in the United States of America

CPSIA Compliance Information: Batch #CW15PK: For Further Information contact Rosen Publishing, New York, New York at 1-800-237-9932

CONTENTS

LIFE IN A TIDE POOL

A tide pool is a pool of salt water left on the beach as waves roll in and out. Tide pools can be small, but they're home to a wide **variety** of living things. All these living things are connected through tide pool food chains.

A food chain is a way of showing how plants and animals depend on each other for food. Each time an animal eats a plant or another animal, another **link** is formed in the food chain. These links show how **energy** is passed from one living thing to another.

Food Chain Fact

Tides are the regular rising and falling of the ocean's surface. They're caused by the pull of the moon and the sun.

RED ALGAE

OYSTER

HERRING GULL

SEA STAR

This food chain is just one example of the many food chains that can exist in a tide pool.

ALWAYS CHANGING

Tide pools are created on rocky shores. As the tide rises, water moves high up on the shore. Then, the tide goes out. When this happens, pockets of water are left behind in the thousands of holes and cracks among the rocks. These natural pools are tide pools.

Tide pools are always changing. The waves that wash in and out with the tides can take animals out of these pools, but they can also bring new creatures into tide pools. Some of these new creatures are plankton, which are important parts of tide pool food chains.

Plankton are very tiny creatures that float in water. Many species, or kinds, of plankton are too small for people to see without a **microscope**.

Food Chain Fact
Waves can wash some unlucky tide pool animals out to sea!

ADAPTING TO A ROCKY LIFE

The living things found in tide pools have all **adapted** to this habitat. A habitat is a place where a plant or animal finds everything it needs to live. The creatures that live in a tide pool have adapted to living in salt water and waves pounding against the rocky shore.

Mussels and oysters stick to the rocks in a tide pool. Crabs are able to hide from the waves in their shell. Eelgrass is a plant that can grow underwater. It can still take in sunlight in tide pools because the water isn't deep.

Food Chain Fact

Blades of eelgrass can grow to be 3 feet (0.9 m) long.

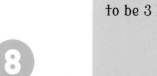

The creatures that live in a tide pool have adapted to a habitat that's always changing as the waves roll in and out.

THE FIRST LINK

Plants and plantlike **organisms** are the first link in tide pool food chains. They get their energy from the sun and then pass that energy to the animals that eat them. Plants and plantlike organisms take in energy from the sun and turn it into food through a **process** called photosynthesis (foh-toh-SIHN-thuh-suhs). During this process, energy from the sun is mixed with water, a gas called carbon dioxide, and other **nutrients** to create a kind of sugar.

One of the most important plantlike organisms in a tide pool is a kind of plankton called phytoplankton.

Food Chain Fact

Phytoplankton float in water, but some tide pool plants, such as sea lettuce, **attach** themselves to the rocks.

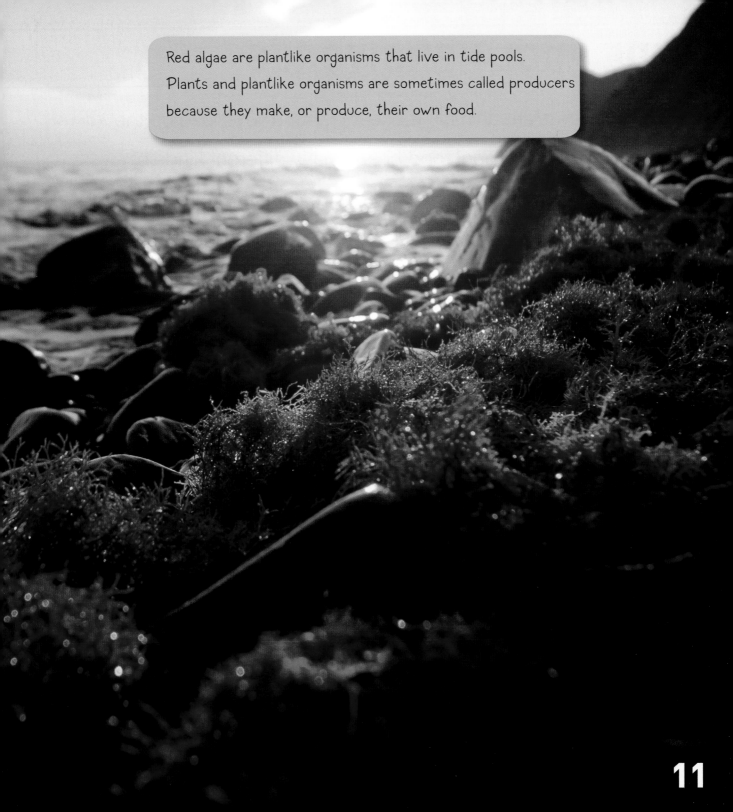

Red algae are plantlike organisms that live in tide pools. Plants and plantlike organisms are sometimes called producers because they make, or produce, their own food.

TIDE POOL HERBIVORES

Herbivores are animals that eat plants. They make up the second link in a tide pool food chain. Many different kinds of herbivores live in tide pools. Limpets are animals with a flat shell that slowly scrape algae off rocks. Some species of sea urchins also eat algae. Shrimp make meals out of the phytoplankton floating in the water.

Food Chain Fact

Although many species of sea urchins are herbivores, some species also eat meat.

Many tide pool herbivores have thick shells to **protect** themselves from predators. They move very slowly, so they can't run from animals that want to eat them!

Limpets are commonly found stuck to the rocks in a tide pool.

DIFFERENT KINDS OF CARNIVORES

Plants aren't the only things that are eaten in a tide pool. Animals often hunt and eat other animals in this habitat. Animals that eat other animals are called carnivores.

Some carnivores, such as the octopus, move through the water to hunt. Others, such as barnacles, sit and wait for their **prey** to drift by. Then, they catch the prey.

Many carnivores in tide pools eat other carnivores. These predators are called secondary carnivores. Oyster drills are secondary carnivores. They use their tongue to drill holes into the shell of their prey, and they suck their prey out through those holes.

Food Chain Fact

Barnacles are small animals covered with hard plates. They can be found stuck to the bottoms of boats or the bodies of other animals, including clams and whales!

Sea stars, or starfish, use their powerful arms to open the shells of other animals. Sea stars turn their stomach outside of their body to eat their prey!

WHAT DO CRABS EAT?

Some animals living in tide pools eat both plants and animals. These animals are called omnivores. Some species of crabs are omnivores.

Even dead plants and animals are food for tide pool animals. Some of the most important parts of tide pool food chains are scavengers. These animals get energy from eating dead animals and plants. Hermit crabs are scavengers. They eat living plants and animals, but they also eat dead plants and animals. A rock crab is another common tide pool scavenger.

Scavengers, such as this hermit crab, are important parts of food chains because they make sure no energy goes to waste.

A hermit crab protects itself from predators by living in the empty shells of other tide pool animals. As the crab grows, it moves on to find a bigger shell.

SUPERSPONGES!

Another kind of animal that lives in a tide pool is the sponge. Sponges are also scavengers, but they don't move around to find food the way crabs do. Instead, sponges take water in through their **pores**. They eat tiny pieces of plant and animal matter in the water.

Sponges also eat bacteria. These tiny creatures are important to food chains because they help to break down dead plants and animals. Creatures that do this are called decomposers. Bacteria are food for many other tide pool animals, including clams and oysters.

bacteria